Computer Solved

Computer Solved

Non-Linear Differential Equations

Joe J. Ettl

Copyright © 2019 by Joe J. Ettl.

ISBN:	Softcover	978-1-7960-2402-9
	eBook	978-1-7960-2401-2

All rights reserved. No part of this book may be reproduced or transmitted in any form or by any means, electronic or mechanical, including photocopying, recording, or by any information storage and retrieval system, without permission in writing from the copyright owner.

The views expressed in this work are solely those of the author and do not necessarily reflect the views of the publisher, and the publisher hereby disclaims any responsibility for them.

Any people depicted in stock imagery provided by Getty Images are models, and such images are being used for illustrative purposes only.
Certain stock imagery © Getty Images.

Print information available on the last page.

Rev. date: 03/26/2019

To order additional copies of this book, contact:
Xlibris
1-888-795-4274
www.Xlibris.com
Orders@Xlibris.com
792465

CONTENTS

Preface vii

Chapter One

1.) $dy/dx = y^2 + x$ 1
2.) $dy/dx = y^2 - x$ 5
3.) $dy/dx = y^2 + x^2$ 8
4.) $dy/dx = y^2 - x^2$ 11
5.) $dy/dx = (y - x)^2$ 14

Chapter Two

1.) $dy/dx = y^3 + x$ 16
2.) $dy/dx = y^3 - x$ 19
3.) $dy/dx = y^3 + x^2$ 22
4.) $dy/dx = y^3 - x^2$ 25
5.) $dy/dx = y^3 + x^3$ 28
6.) $dy/dx = y^3 - x^3$ 31

Chapter Three

1.) $dy/dx = (x + y + x*y)/(x + y - x*y)$ 34
2.) $dy/dx = (x + y - x*y)/(x + y + x*y)$ 38

Chapter Four

1.) $y'' = -y + 0.1*y^3$ 42
2.) $y'' = -y - .1*y^3$ 43
3.) $y'' + \sin y = 0$ 44
4.) $y'' + y' + \sin y = 0$ 46

Conclusions 47

PREFACE

A mid-point approach is used to find the numerical solution of a non-linear differential equation. The t, x, y, z, dy/dx, dy/dt, dx/dt, dz/dx are replaced by T, X+DX/2, Y+DY/2, Z+DZ/2, DY/DX, DY/DT, DX/DT, DZ/DX. The higher orders of the dependent variable are discarded and in some cases both the dependent and the independent variables were discarded.

A scientific pocket computer (1970's version) was used. The enter key continues the next computation. A stop command does not have to be included. All computer programs were written in the old basic language. The graphs were done by hand from the tabular values. A maximum, minimum, inflection, zero, and an asymptote can be found from the graphs.

The rate of convergence can be seen by doubling the number of computations for the Independent variable. If 8, 16, and 32 are used for the same change of 0.05, the rate of 16 computations gives results good to four significant figures (except for asymptotes).

Chapters One and Two deals with dy/dx = y^2 + f (x, y) and dy/dx = y^3 + g (x, y). Chapter Three deals with dy/dx = (f(x,y)) / (g(x,y)). Chapter Four deals with a soft spring, a hard spring, and a pendulum. Each example is worked out so the computer programs can be used.

POCKET COMPUTER PROGRAM ONE

This program is used for dy/dx = y^2 + f (x,y) and dy/dx = y^3 + g (x,y). The B and C will be shown in each example.

10 INPUT "X=";X,"Y=";Y,"M=",M
20 FOR N = 1 TO M STEP 1
30 DX = +.05/M or DX = -.05/M
40 A = X + DX/2
50 B =
60 C =
70 DY = B/C
80 X = X + DX
90 Y = Y + DY
100 NEXT N
110 PRINT"X=";X
120 PRINT Y
120 GO TO 20

POCKET COMPUTER PROGRAM TWO

This program is used for dy/dx = F(x,y)/G(x,y).
    The A, B, C, D, E, and F will be found in each example.
    200 INPUT "X=";X,"Y=";Y,"T=";T,"M=";M
    210 FOR N = 1 TO M STEP 1
    220 DT = +.05/M or -.05/M
    230 A =
    240 B =
    250 C =
    260 D =
    270 E =
    280 F =
    290 DX = (C*E - B*F) / (A*E − B*D)
    300 DY = (A*F − C*D) / (A*E − B*D)
    310 X = X + DX
    320 Y = Y + DY
    330 T = T + DT
    340 NEXT N
    350 PRINT "T=";T
    360 PRINT X
    370 PRINT Y
    380 GO TO 210

POCKET COMPUTER
PROGRAM THREE

This program is used for the second derivative. The A, B, C, D, E, and F will be found in each example.

```
400 INPUT "X=";X,"Y=";Y, "Z="; Z, "M=";M
410 FOR N = 1 TO M STEP 1
420 DX =.05/M or -.05/M
430 A =
440 B =
450 C =
460 D =
470 E =
480 F =
490 DY = (C*E − B*F) / (A*E − B*D)
500 DZ = (A*F − C*D) / (A*E − B*D)
510 X = X + DX
520 Y = Y + DY
530 Z = Z + DZ
540 IF Y < O THEN PRINT "X";X
550 NEXT N
560 PRINT "X=";X
570 PRINT Y
580 GO TO 410
```

Note: The computer must be in the radian mode for the pendulum example.

RATE OF CONVERGENCE

1.) For $dy/dx = y^2 + 1$, a solution is $y = \tan\left(x + \dfrac{\pi}{4}\right)$ and the approximate solution is $DY = (Y^2 + 1) / (1 / DX - Y)$.
   a.) x=0.75, y=28.2383, Y(M=8)=28.2305, Y(M=16)= 28.2363, Y(M=32)=28.2378
   b.) x=0.80, y=-68.4797, Y(M=8)=-68.5286, Y(M=16)=-68.4919, Y(M=32)=-68.4827

2.) For $dy/dx = y^3$, a solution is $y^2 = 1/(1 - 2*x)$ and the approximate solution is
   $DY = (Y^3) / (1 / DX - 1.5 * Y^2)$.
   a.) x=0.45, y=3.1623, Y(M=8)=3.1609 Y(M=16)=3.1619, Y(M=32)=3.1622
   b.) x=.50, y is undefined, Y(M=8)=23.505. Y(M=16)=33.209, Y(M=32)=46.942
   c.) x=.55, y has no real value, Y(M=8)=2.7770, Y(M=16)= -6.7970, Y(M=32)= 9.9.9632
   d.) There are no converging values after x = 0.50

3.) For $dy/dx = y^4 + f(x, y)$, the computer values were not reliable.

EXAMPLE ONE

1.) dy/dx = y^2 + x
2.) A = X + DX / 2
3.) DY/DX does not equal (Y + DY/2)^2 + A
4.) DY/DX = Y^2 + Y*DY + A
5.) DY/DX − Y*DY = Y*2 + A
6.) (1/DX − Y)*DY = Y*2 + A
7.) DY = (Y*2 + A) / (1 / DX − Y)
8.) B = Y^2 + A and C = 1/DX − Y
9.) As $x \to -\infty$, $Y \to -\sqrt{-x}$

dy/dx = y^2 + x

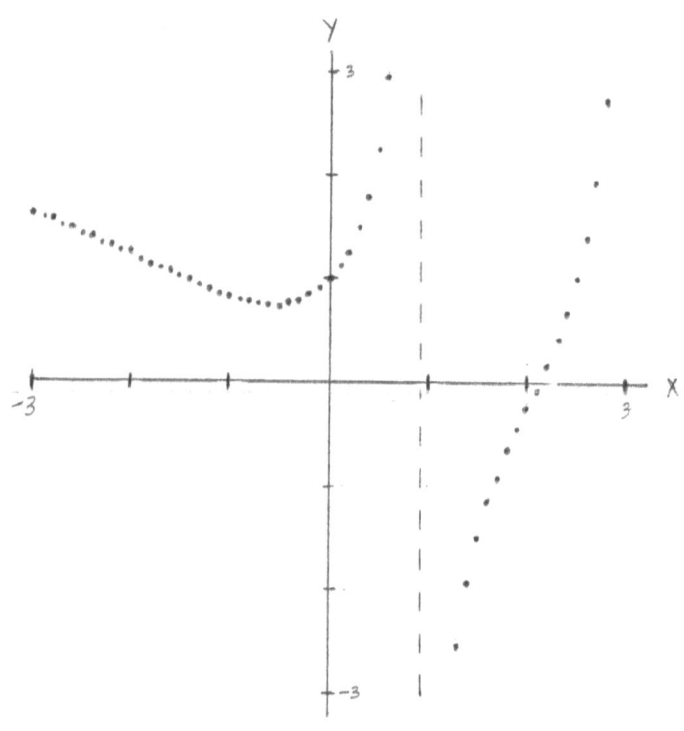

dy/dx = y^2 + x for x=0 and y = 0

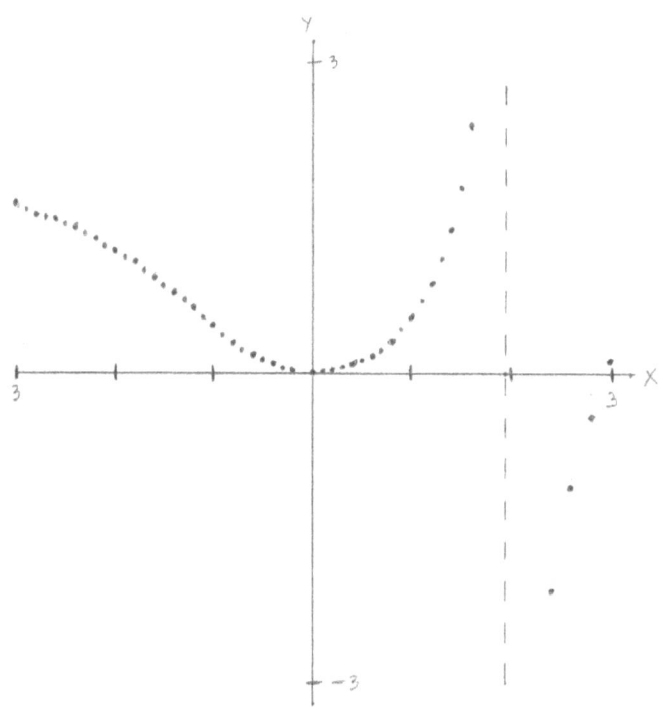

dy/dx = y^2 + x for x=0 and y=0

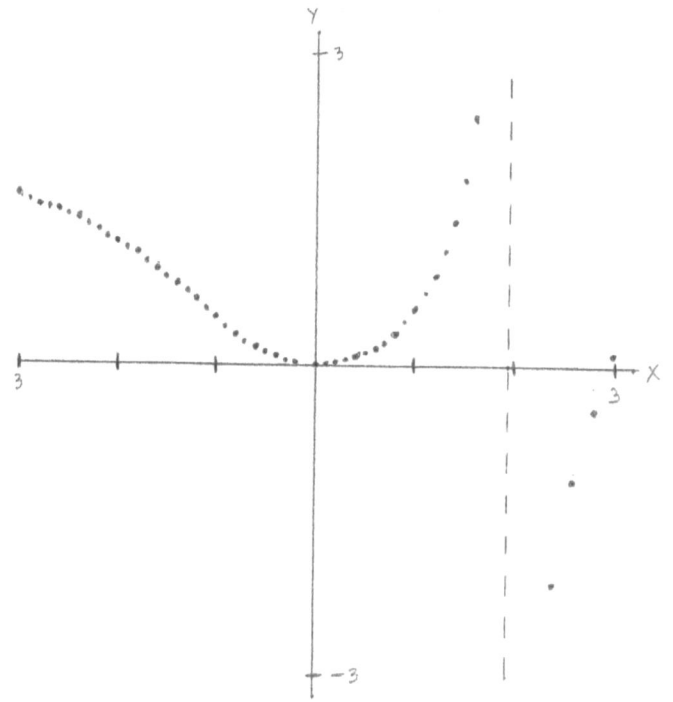

EXAMPLE TWO

1.) dy/dx = y^2 – x
2.) A = X + DX / 2
3.) DY / DX does not equal (Y + DY / 2)^2 –A
4.) DY / DX = Y^2 + Y* DY – A
5.) DY / DX – Y* DY = Y^2 -A
6.) (1 / DX – Y) = Y^2 – A
7.) DY = (Y^2 – A) / (1 DX – Y)
8.) B = Y^2 – A and C = 1 / DX-Y
9.) *As* $x \to +\infty$, $Y \to -\sqrt{x}$

$dy/dx = y^2 - x$

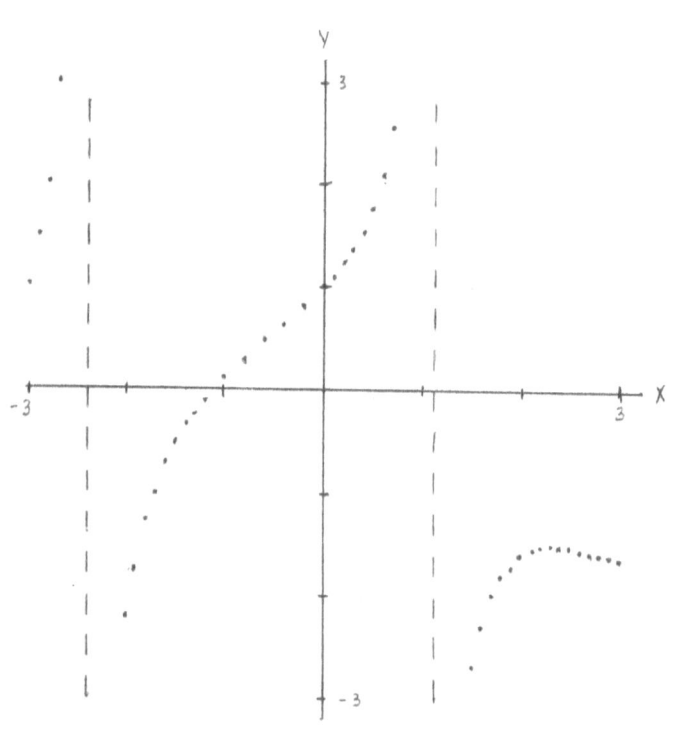

dy/dx = y^2 − x

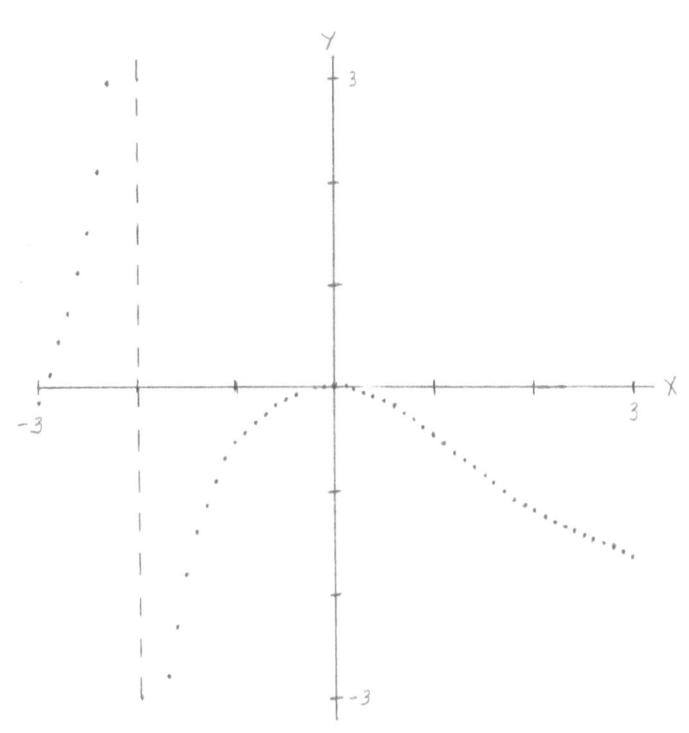

EXAMPLE THREE

1.) dy/dx = y^2 + x^2
2.) A = X + DX / 2
3.) DY/DX does not equal (Y + DY/2) ^2 + A^ 2
4.) DY/DX = Y^2 + Y * DY + A ^ 2
5.) DY/DX – Y * DY = Y^2 + A^ 2
6.) (1 / DX – Y) * DY = Y ^ 2 + A ^ 2
7.) DY = (Y^2 + A ^ 2) / (1 DX – Y)
8.) B = Y ^ 2 + A ^ 2 and C = 1 / DX — Y

dy/dx = x^2 + y^2

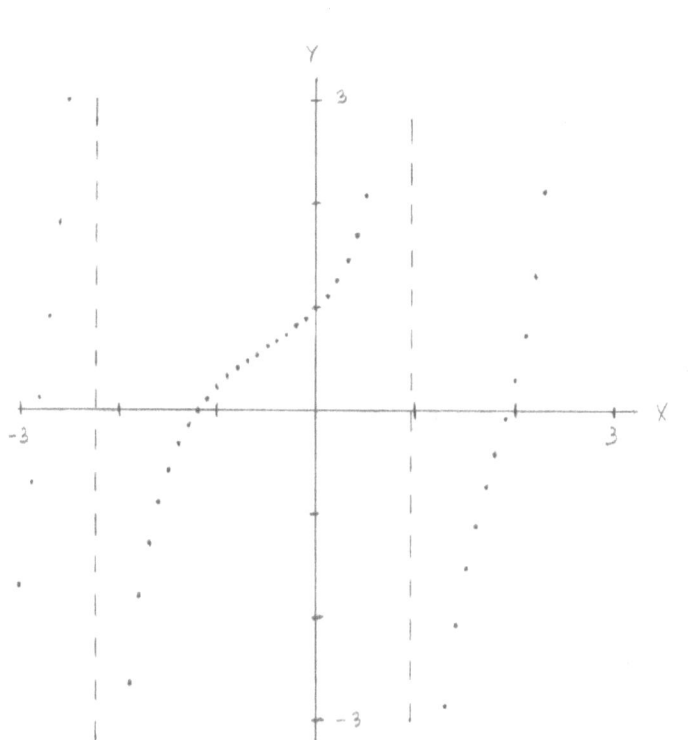

$dy/dx = x^2 + y^2$

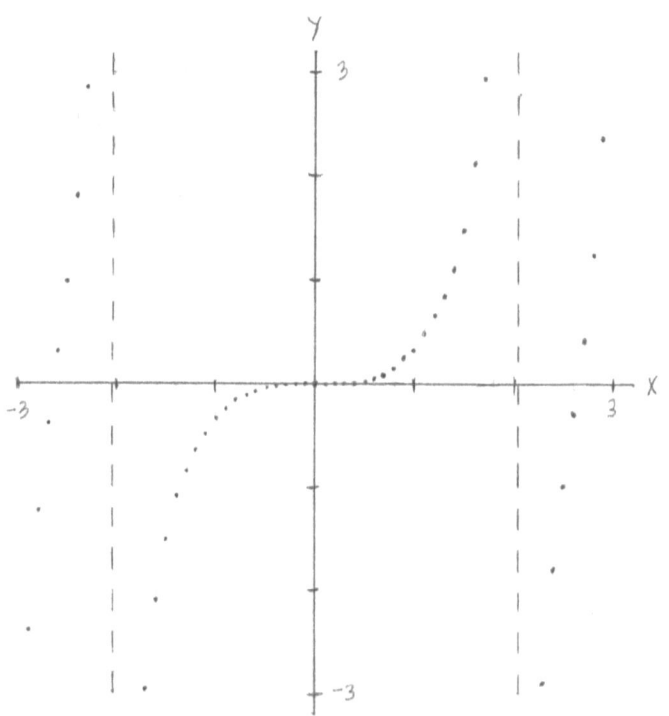

EXAMPLE FOUR

1.) $dy/dx = y^2 - x^2$
2.) $A = X + DX/2$
3.) DY/DX does not equal $(Y + DY/2)^2 - A^2$
4.) $DY/DX = Y^2 + Y*DY - A^2$
5.) $DY/DX - Y*DY = Y^2 - A^2$
6.) $(1/DX - Y)*DY = Y^2 - A^2$
7.) $DY = (Y^2 - A^2) / (1\ DX - Y)$
8.) $B = Y^2 - A^2$ and $C = 1/DX - Y$
9.) $Y \rightarrow -x$ as $x \rightarrow \infty$ and as $x \rightarrow -\infty$

$dy/dx = y^2 - x^2$

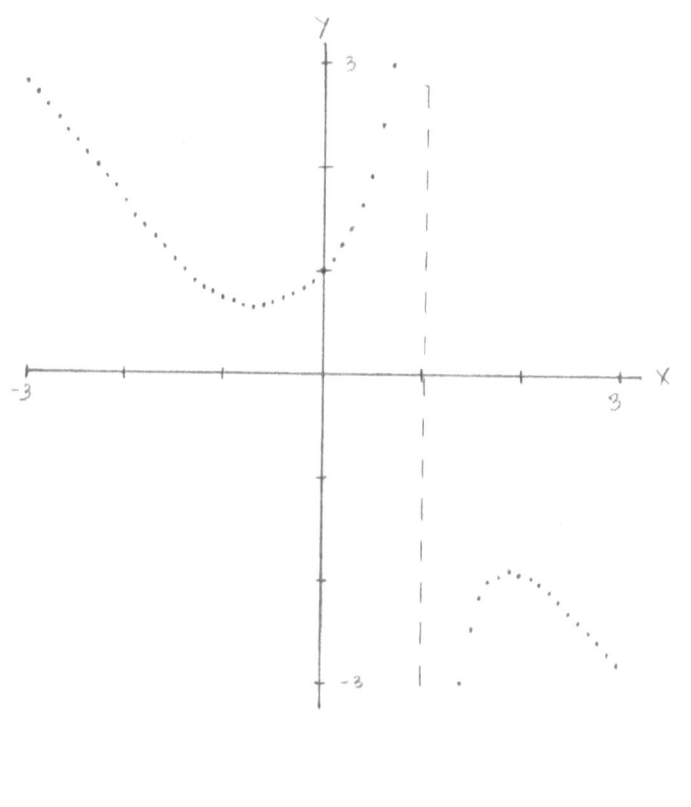

$$dy/dx = y^2 - x^2$$

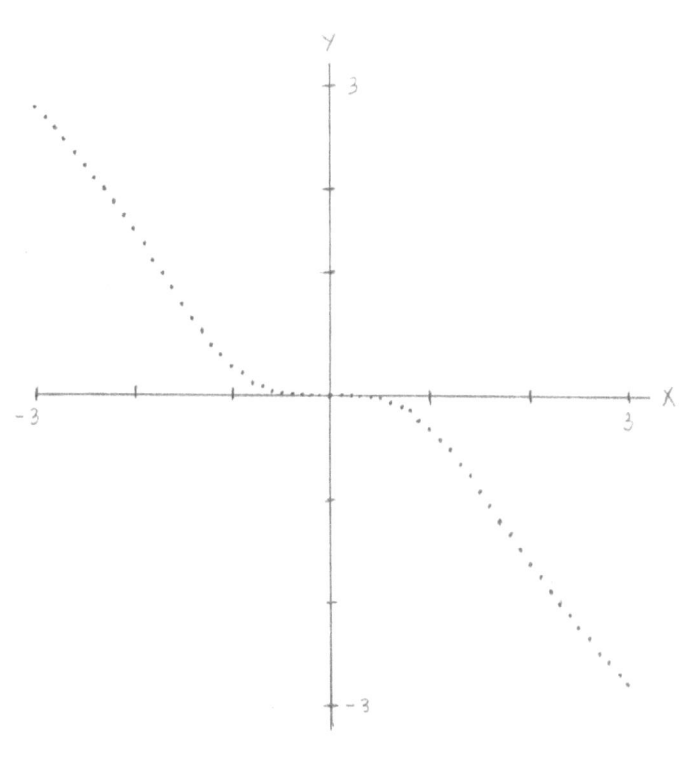

EXAMPLE FIVE

1.) dy/dx = (y − x) ^2
2.) A = X + DX / 2
3.) DY/DX does not equal ((Y + DY/2) −A)^ 2
4.) DY/DX does not equal (Y + DY/2)^ 2 - 2 * A * (Y + DY /2) + A ^ 2
5.) DY/DX = Y ^ 2 + Y * DY − 2 * A * Y - A * DY + A^ 2
6.) DY/DX − Y * DY + A * DY = (Y − A) ^ 2
7.) (1 / DX − (Y − A))* DY = (Y − A) ^ 2
8.) DY = ((Y − A) ^ 2) / (1 / DX − (Y − A))
9.) B = (Y − A) ^ 2 and C = 1 / DX − (Y − A)
10.) For the starting points (0,0) and (0, 2), the results are:

As $x \to +\infty$, $Y \to x - 1$
As $x \to -\infty$, $Y \to x + 1$

The point (0, 2) has a vertical asymptote between x = 0.50 and x = 0.55.

dy/dx = (y − x) ^ 2

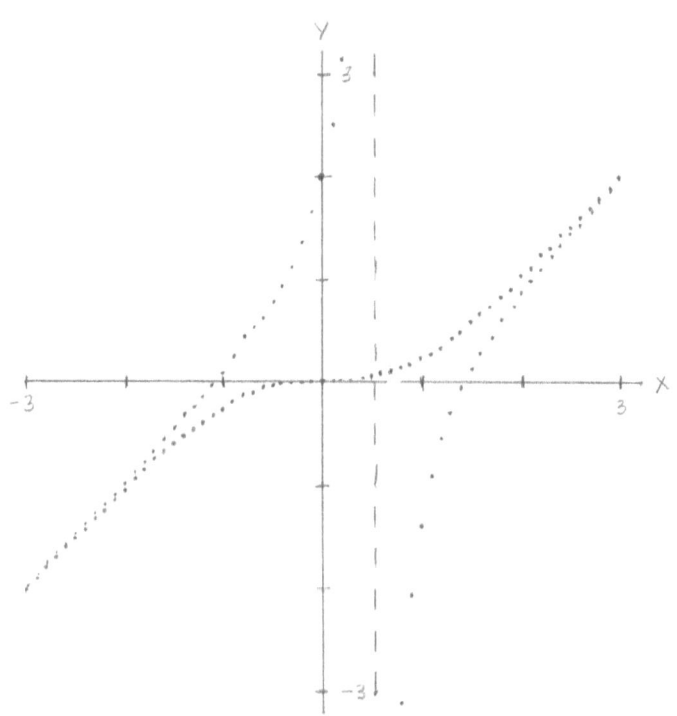

EXAMPLE ONE

1.) dy/dx = y^3 + x
2.) A = X + DX/2
3.) DY/DX » (Y + DY/2)^3 + A
4.) DY/DX = Y^3 + 1.5*Y^2*DY + A
5.) (1/DX -1.5*Y^2) * DY = Y^3 + A
6.) DY = (Y^3 + A) / (1/DX − 1.5*Y^2)
7.) B = Y^3 + A and C = 1/DX − 1.5^2
8.) There are no converging values of Y beyond the vertical asymptote.
9.) As $x \to -\infty$, then $Y \to \sqrt[3]{-X}$
10.) If the quadratic equation DY/DX = A + Y^3 + 3*Y^2*(DY/2) +3*(DY/2)^2 is used, then the discriminate will become negative. This gives a better approximation to the independent value that becomes a complex number. The drawback is the starting point of (0,0) cannot be used. However, this can be done using the program above and then using the values of X and Y (not zero) in a different program to solve the quadratic equation. This will not be shown.

dy/dx = y^3+x

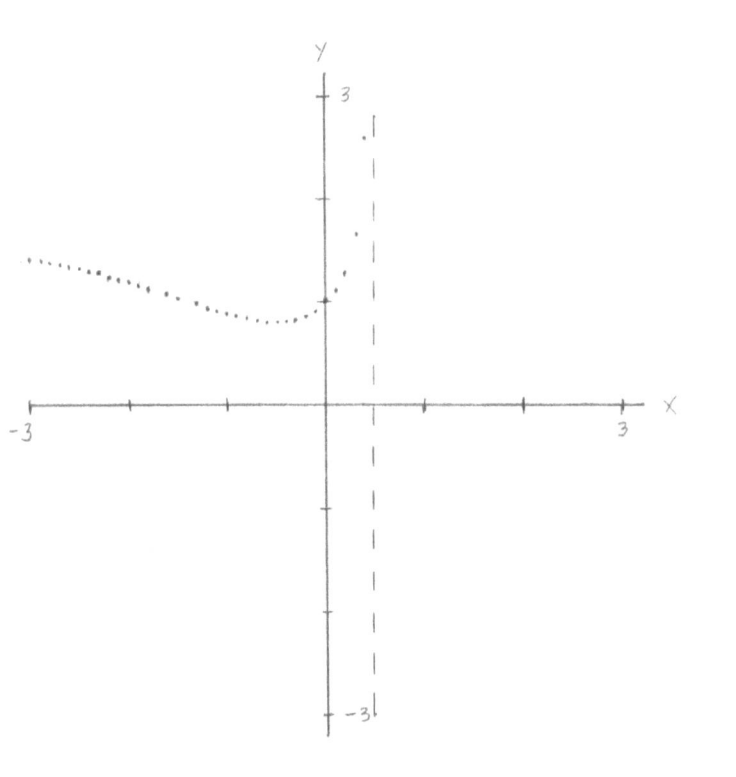

dy/dx = y^3 + x

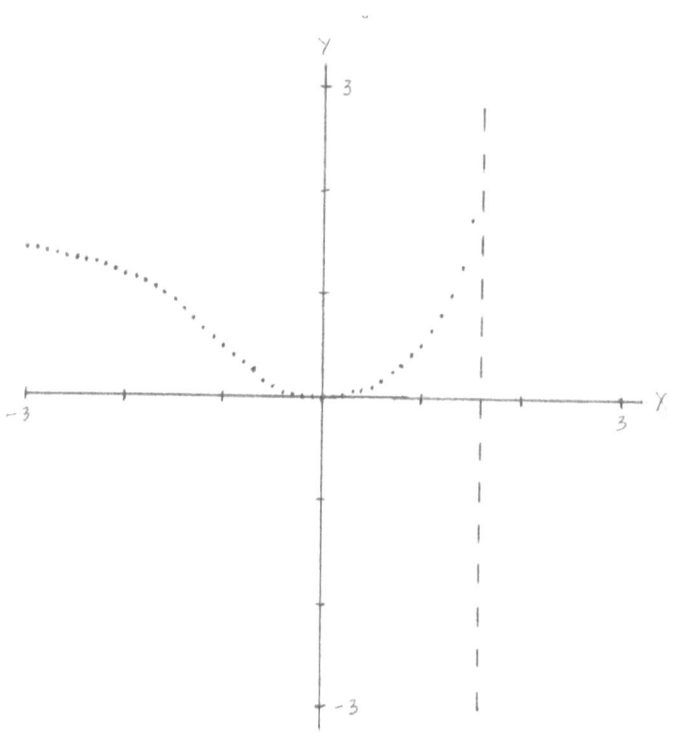

EXAMPLE TWO

1.) dy/dx = y^3 − x
2.) A = X + DX/2
3.) DY/DX » (Y + DY/2)^3 − A
4.) DY/DX = Y^3 + 1.5*Y^2*DY − A
5.) (1/DX − 1.5*Y^2)*DY = Y^3 − A
6.) DY = (Y^3 − A)/ (1/DX − 1.5*Y^2)
7.) B = Y^3 − A and C = 1/DX − 1.5*Y^2
8.) There are no converging values beyond the vertical asymptote.
9.) $As\ X \to -\infty,\ Y \to \sqrt[3]{X}$

dy/dx = y^3 −x

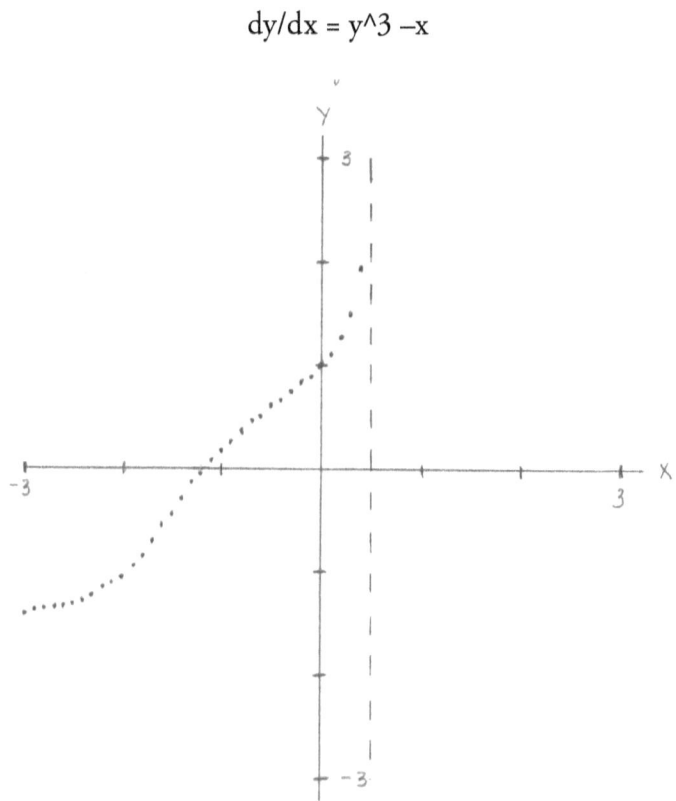

dy/dx = y^3 −x

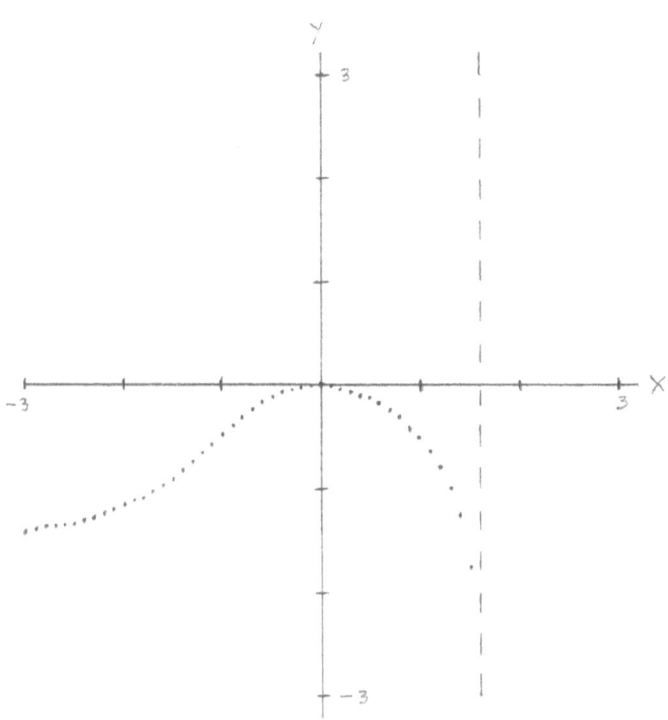

EXAMPLE THREE

1.) $dy/dx = y^3 + x^2$
2.) $A = X + DX/2$
3.) $DY/DX \approx (Y + DY/2)^3 + A^2$
4.) $DY/DX = Y^3 + 1.5*Y^2 * DY + A^2$
5.) $(1/DX - 1.5*Y^2)* DY = Y^3 + A^2$
6.) $DY = (Y^3 + A^2)/(1/DX - 1.5*Y^2)$
7.) $B = Y^3 + A^2$ and $C = 1/DX - 1.5*Y^2$
8.) There are no converging values of Y beyond the vertical asymptote
9.) As $X \to -\infty$, $Y \to \sqrt[3]{-X^2}$

$$\frac{dy}{dx} = y^3 + x^2$$

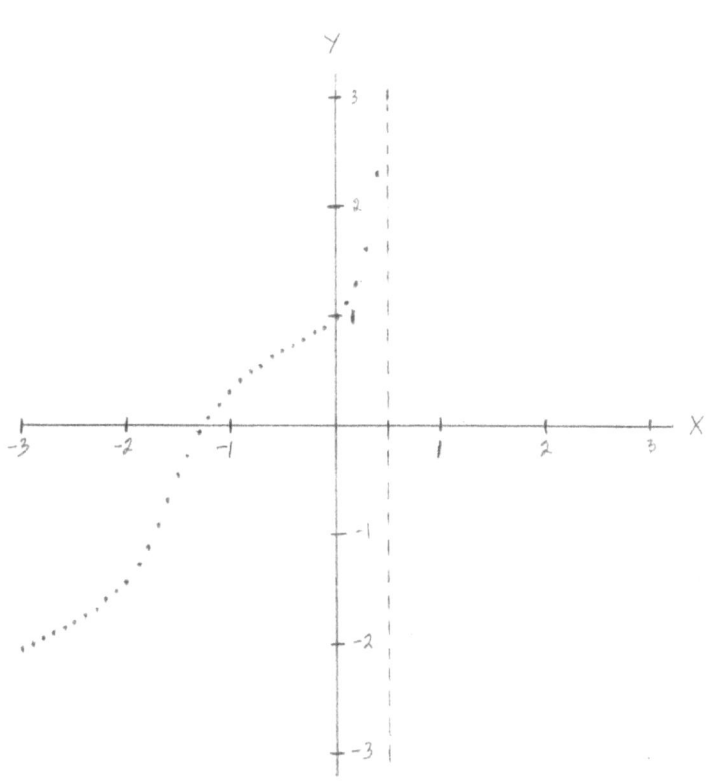

$$\frac{dy}{dx} = y\wedge 3 + x\wedge 2$$

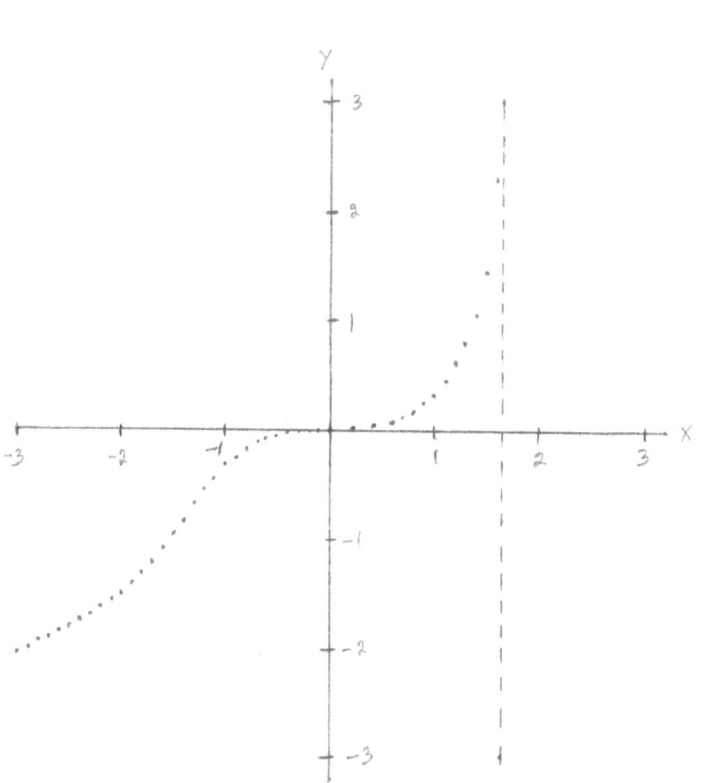

EXAMPLE FOUR

1.) dy/dx = y^3 − x^2
2.) A = X + DX/2
3.) DY/DX ≈ (Y + DY/2)^3 − A^2
4.) DY/DX = Y^3 + 1.5*Y^2 − A^2
5.) (1/DX − 1.5*Y^2)*DY = Y^3 − A^2
6.) DY = (Y^3 − A^2)/ (1/DX − 1.5*Y^2)
7.) B = Y^3 − A^2 and C = 1/DX − 1.5*Y^2
8.) There are no converging values beyond the vertical asymptote
9.) *As* $X \to -\infty$, $Y \to \sqrt[3]{X^{\wedge}2}$

$$\frac{dy}{dx} = y^3 - x^2$$

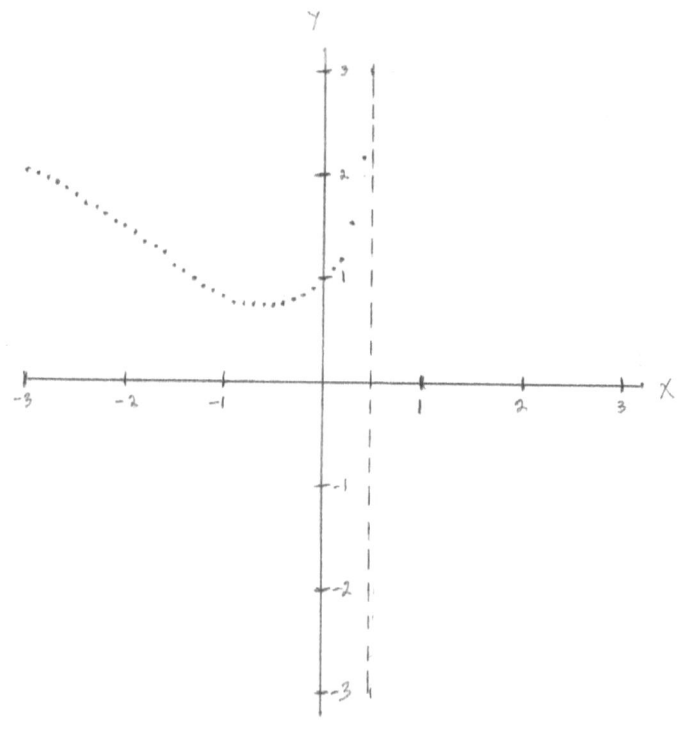

$$\frac{dy}{dx} = y\wedge 3 - x \wedge 2$$

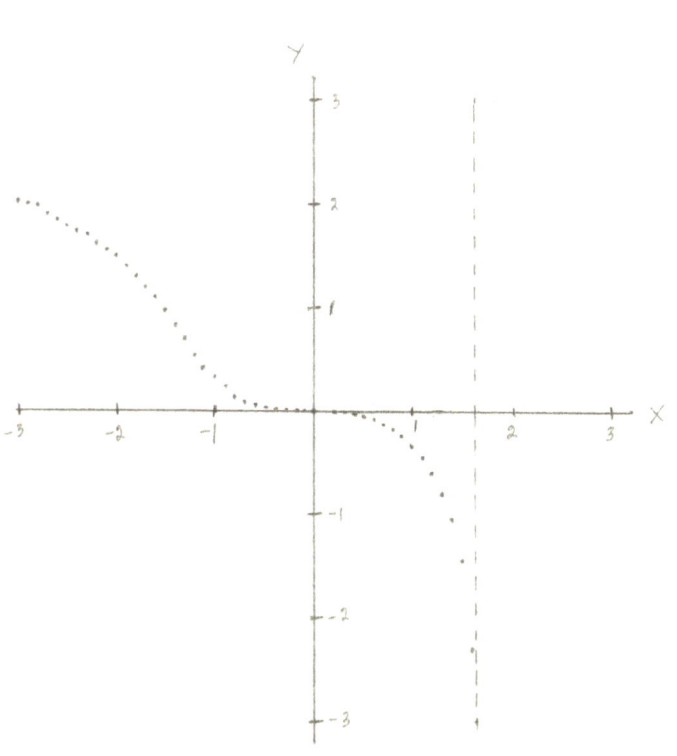

EXAMPLE FIVE

1.) dy/dx = y^3 + x^3
2.) A = X + DX/2
3.) DY/DX ≈ (Y + DY/2)^3 + A^3
4.) DY/DX = Y^3 + 1.5*Y^2*DY + A^3
5.) (1/DX-1.5*Y^2)*DY = Y^3 + A^3
6.) DY = (Y^3 + A^3) / (1/DX -1.5*Y^2)
7.) B = Y^3 + A^3 and C = 1/DX − 1.5*Y^2
8.) There are no converging values of Y beyond the vertical asymptote
9.) *As $X \to -\infty$, $Y \to -X$*

$$\frac{dy}{dx} = y\wedge 3 + x\wedge 3$$

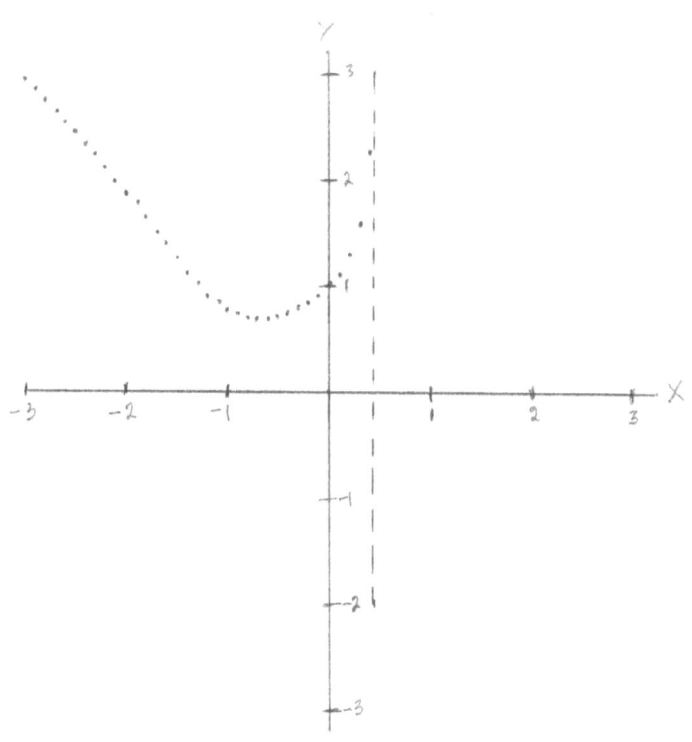

$$\frac{dy}{dx} = y^3 + x^3$$

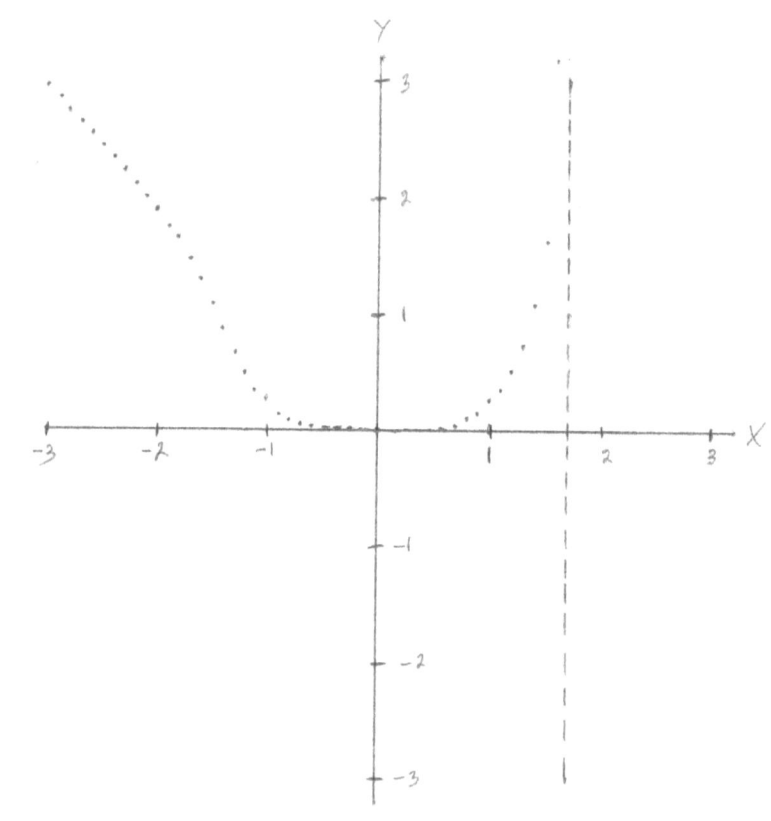

EXAMPLE SIX

1.) dy/dx = y^3 − x^3
2.) A = X + DX/2
3.) DY/DX ≈ (Y + DY/2)^3 − A^3
4.) DY/DX = Y^3 + 1.5*Y^2*DY − A^3
5.) (1/DX - 1.5*Y^2)*DY = Y^3 − A^3
6.) DY = (Y^3 - X^3)/(1/DX − 1.5*Y^2)
7.) B = Y^3 − X^3 and C = 1/DX − 1.5*Y^2
8.) There are no converging values of Y beyond the vertical asymptote
9.) *As $X \to -\infty$, then $Y \to X$*

$$\frac{dy}{dx} = y\wedge 3 - x\wedge 3$$

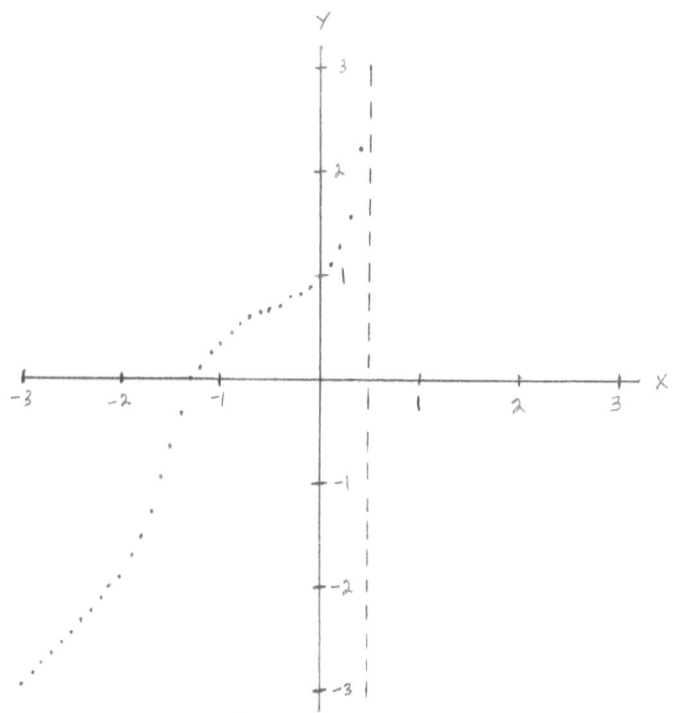

$$\frac{dy}{dx} = y^3 - x^3$$

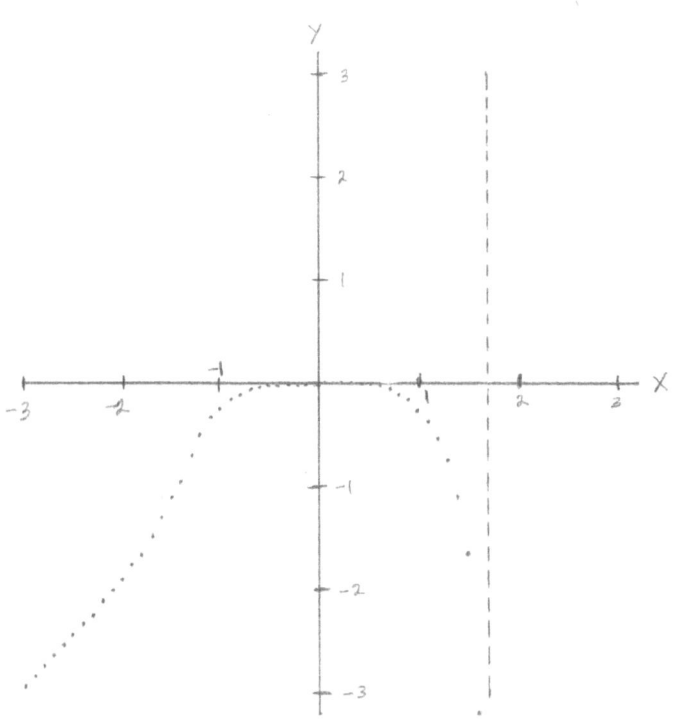

EXAMPLE ONE

1.) $dy/dx = (x + y + x*y) / (x + y - x*y)$
2.) $dy/dt = x + y + x*y$ and $dx/dt = x + y - x*y$
3.) $DY/DT \approx (X + DX/2) + (Y + DY/2) + (X + DX/2)*(Y + DY/2)$
   and
   $DX/DT \approx (X + DX/2) + (Y + DY/2) - (X + DX/2)*(Y + DY/2)$
4.) The terms of $(DX/2)* DY/2)$ for DX/DT and DX/DT are discarded.
   $DY/DT = X + .5*DX + Y + .5*DY + X + Y + .5*X*DY + .5*Y*DX$
   and
   $DX/DT = X + .5*DX + Y + .5*DY - X*Y - .5*X*Y - .5*X*DY - .5*Y*DX$
5.) The like terms are arranged for DY/DT and DX/DT
   $(-.5 - .5*Y)*DX + (1/DT - .5 - .5*X)*DY = X + Y + X*Y$ for DY/DT
   $(1/DT - .5 + .5*Y)*DX + (-.5 + .5*X)*DY = X + Y - X*Y$ for DX/DT
6.) Use pocket computer program two.
   $A = -.5 - .5*Y, B = 1/DT -.5 - .5*X, C = X + Y + X*Y$
   $D = 1/DT -.5 + .5*Y, E = -.5 + .5*X, F = X + Y - X*Y$

$dy/dt = x + y + xy$

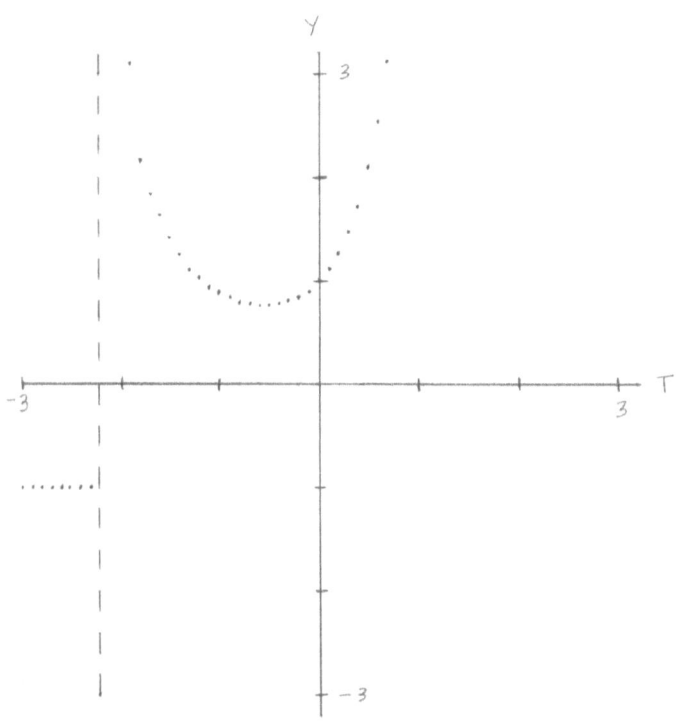

$dx/dt = x + y - xy$

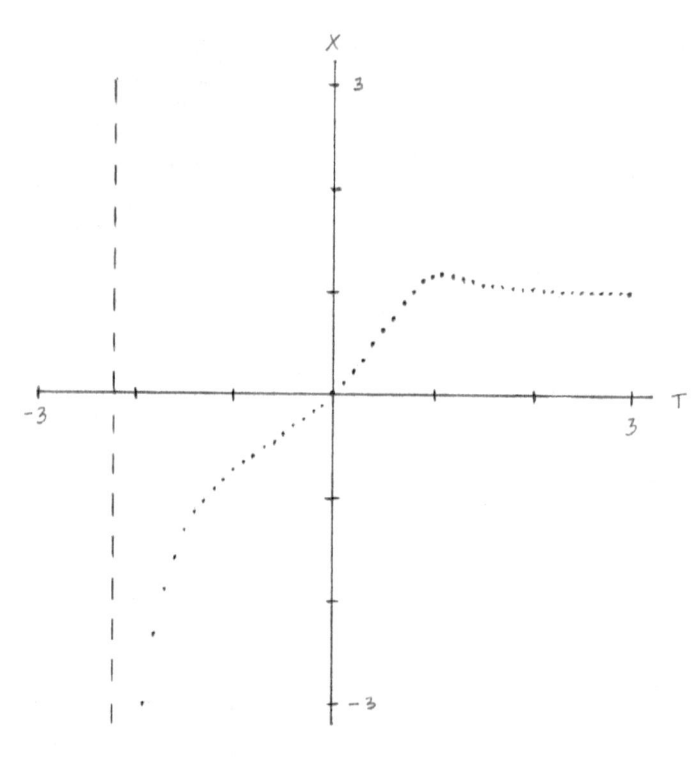

dx/dx = (x + y + xy) / (x + y − xy)

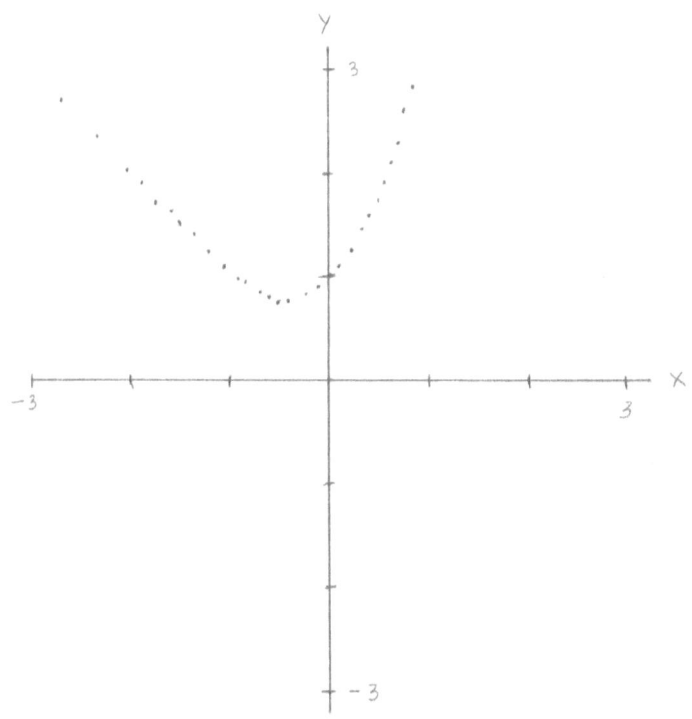

EXAMPLE TWO

1.) $dy/dx = (x + y - x*y)/(x + y + x*y)$
2.) $dy/dt = x + y - x*y$ and $dx/dt = x + y + x*y$
3.) Follow exercise one to obtain:
    $A = -.5 + .5*Y$, $B = 1/DT -.5 + .5*X$, $C = X + Y - X*Y$
    $D = 1/DT - .5 -.5*Y$, $E = -.5 - .5*X$, $F = X + Y + X*Y$
4.) The numerical solution of $dy/dx = (y^3 + x^3)/(y^3 - x^3)$ is easily done using the procedure of exercises one and two.
    If $y/x = u$ where $(1,0)$ is the starting point or $x/y = u$ where $(0,1)$ is the starting point, a complex non-closed solution can be found. This is not a non-linear differential equation.

dy/dt = x + y − xy

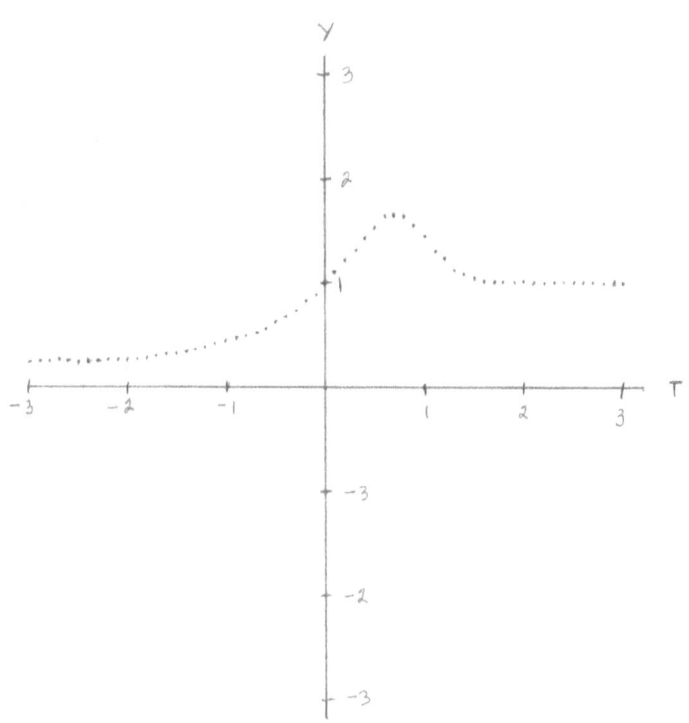

dy/dt= x + y + xy

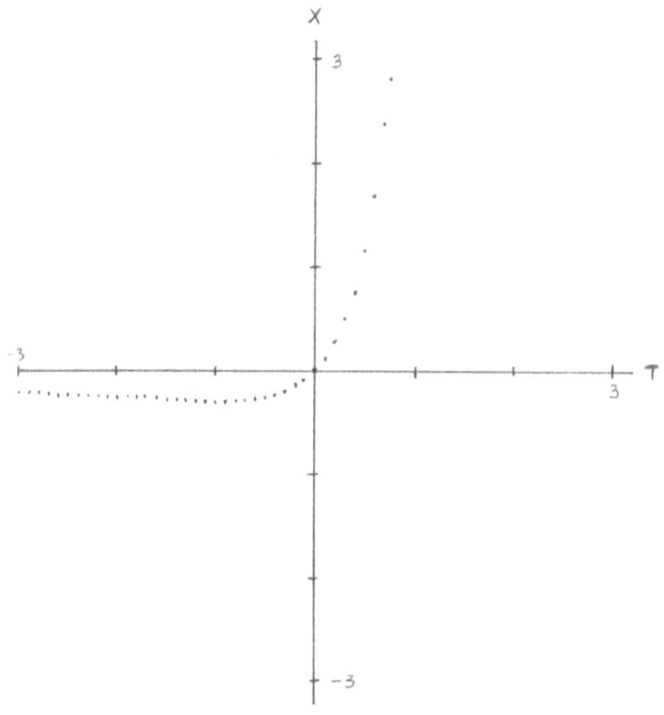

$dy/dx = (x + y - xy) / (x + y + xy)$

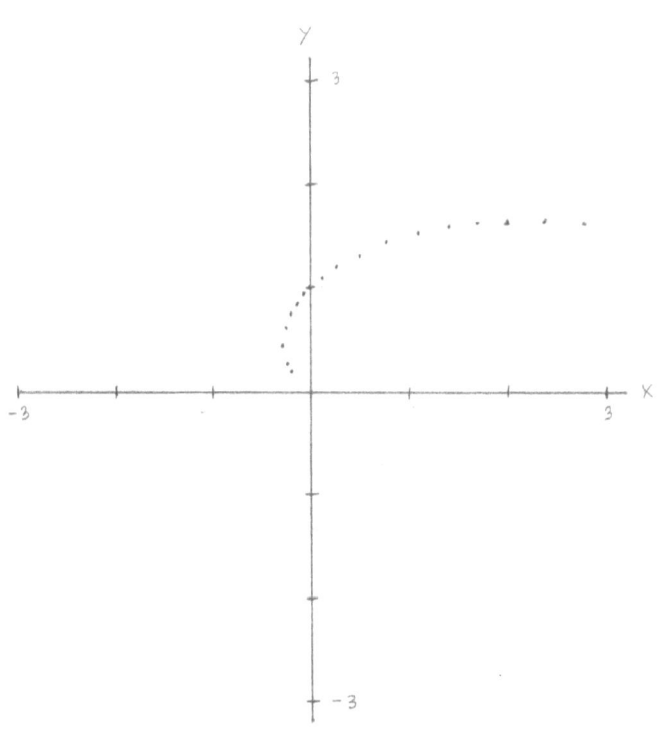

EXAMPLE ONE

1.) $y'' = -y + 0.1*y^3$, $z = dy/dx$, $dz/dx = y''$ represents a hard spring.
2.) $dy/dx = z$ and $dz/dx = -y + .1*y^3$
3.) For $dy/dx = z$, then $DY/DX = Z + DZ/2$ or $(1/DX)*DY + (-.5)*DZ = Z$.
4.) For $dz = -y + .1*y^3$, then $DZ/DX \approx - (Y + DY/2) + (.1)*(Y+DY/2)^3$ or $DZ/DX = -Y - .5*DY + (.1)*Y^3 + (.15)*Y^2*DY$ or
$(.5 - .5*Y^2)*DY + (1/DX)*DZ = -Y + .1*Y^3$
5.) $A = 1/DX$, $B = -.5$, $C = Z$, $D = .5 - .15*Y^2$, $E = 1/DX$, $F = -Y + .1*Y^3$.
Use pocket computer program three.
6.) A zero is bounded for the first quarter of a complete cycle.
One complete cycle is bounded between $4*X = 6.53$ and $4*X = 6.54$.
7.) For $y'' = -y$, one complete cycle is $4*X = 2*\pi$ or $4*X = 6.28$.
This is a perfect extension spring.

$y'' = -y + .1\, y^3$ (Hard Extension Spring)

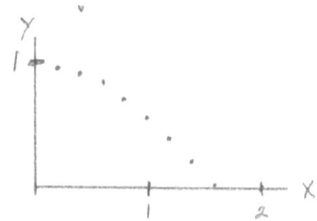

For $Y < 0$, then $x = 1.634375$

EXAMPLE TWO

1.) $y'' = -y - .1*y^3$, $z = dy/dx$, $dz/dx = y''$ represents a soft spring.
2.) Use exercise one as a guide and the pocket computer program three.
   $A = 1/DX$, $B = -.5$, $C = Z$, $D = .5 + .15*Y^2$, $E = 1/DX$, $F = -Y - .1*Y^3$
3.) A zero is bounded for he first quarter of a complete cycle. One complete cycle is bounded between $4*X = 6.05$ and $4*X = 6.06$.
4.) The prefect extension spring cycle is approximately 6.28.

$$y'' = -y - .1\, y^3 \text{ (Soft Extension Spring)}$$

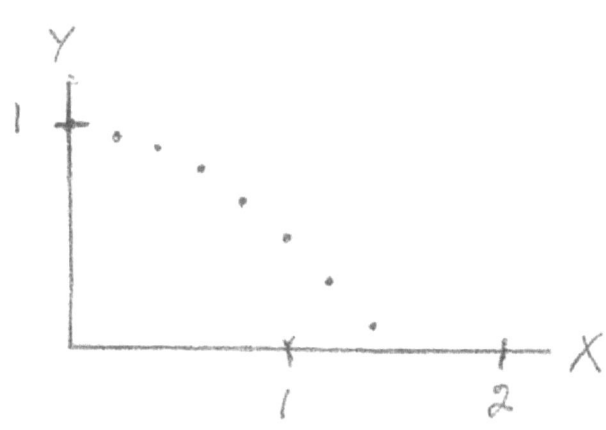

EXAMPLE THREE

1.) For a pendulum $q" = -(g * sinq)/L$ where $q"$ is the second derivative of q *with* respect to time. Let $y = q$ *be measured* in radians and $x\wedge 2 = g*t/L$, then the new equation transforms to $y" = -\sin y$ where $y"$ is the second derivative with respect to x.
2.) Let $dy/dx = z$, then $dz/dx = -\sin y$.
3.) For $dy/dx = z$, then $DY/DX = Z + DZ/2$ or $(1/DX)*DY + (-.5)*DZ = Z$. radians
4.) For $dz/dx = -\sin y$, then $DZ/DX \approx -\sin(Y + DY/2)$ or $DZ/DX \approx - (\sin Y*\cos(DY/2) + \cos Y*\sin(DX/2))$
5.) For small values of DY then $DZ/DX = -\sin Y - \cos Y*(DY/2)$ or $(\cos Y/2)*DY + (1/DX)*DZ = -\sin Y$.
6.) Use pocket computer program three (put in radian mode) $A = 1/DX$, $B = -.5$, $C = Z$, $D = (\cos Y)/2$, $E = 1/DX$, $F = -\sin Y$.
7.) For $X = 0$, $Y = 1$ radian, $Z = 0$, $DX = .05/16$, one complete cycle is
bounded by $4*X = 6.675$ and $4*X = 6.700$.
For $X = 0$, $Y = 2$ radians, $Z = 0$, $DX = .05/16$, one complete cycle is
bounded by $4*X = 8.338$ and $4*X = 8.350$.
For $X = 0$, $Y = 3$ radians, $Z = 0$, $DX = .05/16$, one complete cycle is
bounded by $4*X = 16.150$ and $4*X = 16.163$.
For $X = 0$, $Y = p$ radians, $Z = 0$, $DX = .05/16$, Y is equal to p for all positive values of X.

$y'' = -\sin y$

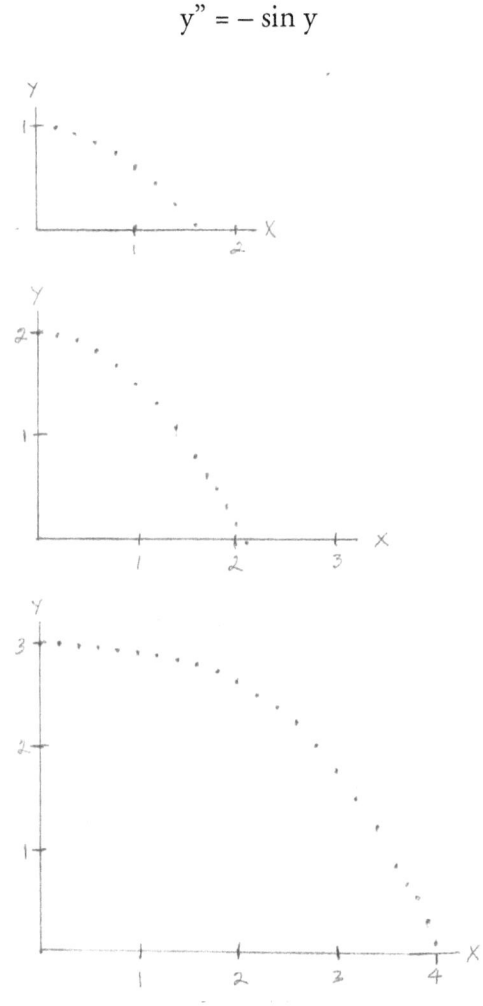

EXAMPLE FOUR

1.) For a pendulum with a linear dampener, the differential equation is $y'' + y' + \sin y = 0$ or $y'' = y' - \sin y$.
2.) Let $dy/dx = z$, then $dz/dx = y''$ or $dz/dx = -z - \sin y$
3.) For dy/dx, then $DY/DX = Z + DZ/2$ or $(1/DX)*DY + (-.5)*DZ = Z$
4.) For $dz/dx = -z - \sin y$ and small values of y, then $DZ/DX = -Z - DZ/2 - \sin Y - (\cos Y)*(DY/2)$ or $((\cos Y)/2)*DY + (1/DX + .5)*DZ = -Z - \sin Y$.
5.) See exercise three (put in radian mode)
 $A = 1/DX$, $B = -.5$, $C = Z$, $D = (\cos Y)/2$, $E = 1/DX + .5$, $F = -Z - \sin Y$
6.) The results for $y'' = -z - \sin y$ are as follows:
 The starting point is $X = 0$, $Y = 1$, $Z = 0$, $DX = .05/16$.
 Zero between (2.60,0.004) and ((2.65,-0.009)
 A minimum point for (3.85,-0.149)
 A zero at (6.25,0.000) and a maximum point at (7.45,0.024)
7.) The results for $y'' = -\sin y$ are as follows:
 The starting point is $X = 0$, $Y = 1$, $Z = 0$, $DX = .05/16$
 A zero between (1.65, 0.024) and (1.70, -0.024)
 A minimum point for ((3.35, -1)
 A zero is between (5.00, -0.024) and (5.05, 0.024)
 A maximum point at (6.70, 1.000)

CONCLUSIONS

1.) DX is equal to dx.
2.) The points were plotted in all graphs to show the changes in the dependent variable for an equal change in the independent variable.
3.) The mid-point approach is easier to use than the Taylor's series for a function of two variables.
4.) The forms of $dy/dx = y^4 + f(x,y)$ and $dy/dx = y^5 + f(x,y)$ were not included.

Reference one is Ettl, Joseph J. "An Alternate Increment Method (AIM)" NYSMATYC Journal, Spring 1970.

Reference two is Ettl, Joseph J. "AIM Predicts Asymptotic Behavior" NYSMATYC Journal, Fall 1970.